A Three-Box Warre Beehive with Stand, Quilt & Roof

You CAN build this!

How to Build a Warre Beehive
Copyright 2020
NHWW
Northwood, New Hampshire

How to Build a Warre Beehive

By W. Todd Abernathy

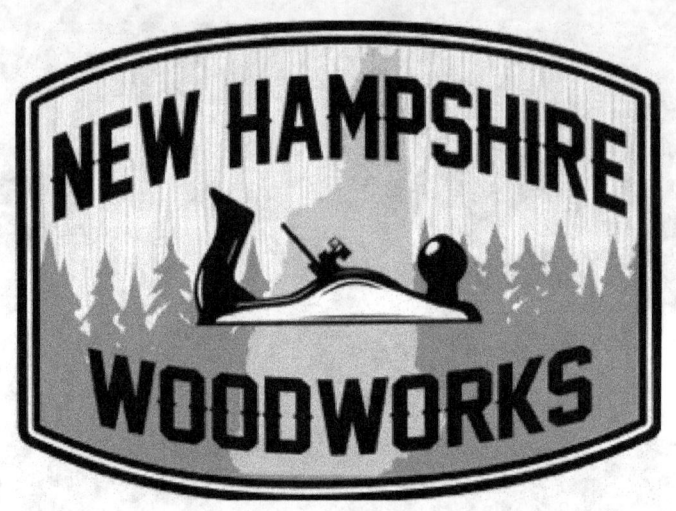

A Homesteading 'How To' Book From

www.UnexplainedUnderfootObjects.com
&
www.NewHampshireWoodWorks.com

Other Books in the Homesteading 'How To' Series:

How to Build a Langstroth Beehive
Available on Amazon.com

Born in the golden age of apiary sciences, Emile Warre was a French monk who pondered the effectiveness of modern beehives. His bee-yard reportedly contained over 800 hives of varying design, each as unique as the genetics held within.

After much observation and experimentation, Abbe Warre devised a simple, inexpensive, productive, and quite ingenious way of caring for bees that best mimicked their natural environment. His idea?

Leave the bees alone.

His design, originally called the *People's Hive*, is a non-invasive methodology of apiary management that encourages bees to go about their business with little interference. With his *la Ruche Populaire*, keeper manipulation is limited to a few times each year rather than every week or two.

Warre hives are a modified top bar design, meaning that the bees build their own fresh comb in each new box along top bar rungs. Rather than line the frames length-wise in a traditional top bar fashion, Warre stacked the frames upwards.

His concept was that rather than re-using manufactured comb that tended to collect pesticides and disease, the ready comb is simply cut and processed from the top bars in only the top box, where bees tend to store excess honey.

New, comb-free boxes and top bars are added from the bottom, continuing the natural cycle. As the bees build new comb at the bottom, they transfer their brood and food stores operations below, using the upper comb for excess honey storage. With this, the hives cycles in a natural fashion.

What could be simpler for both the bees and the beekeeper?

Abbe Eloi Francois Emile Warre (1867-1951)

In this Homesteading 'How To' book, you will learn how to build one of the simplest and most perfectly-engineered beehives known to beekeepers the world over. The materials are inexpensive, the skills required are novice-level, and in the end, you will have something beautiful that your bees will flourish in.

List of Tools

- Simple squaring jig (plans on the next page)
- Table saw with edge guide fence
- Dado blade (optional)
- Miter saw
- Handheld drill (2 for efficiency)
- 5/64" drill bit (countersink optional)
- Four (4) 24" F-bar clamps
- Jig saw (optional)
- Staple gun

Lumber for Squaring Jig

- 4ft x 1" x 2" oak board
- 4ft x 1" x 3ft sheet of plywood

Lumber for a 3-box Warre Hive

- Three (3) 12ft x ¾" x 12" select pine boards
- 3/8" or ½" exterior plywood to be cut to 15 3/8" x 13 ¾"
- 1 4ft 4"x4" post, trimmed to 2.5"x2.5"

Hardware

- One (1) box of #8 1-5/8" exterior deck screws
- One (1) box of #6 1-1/4" steel screws
- Cloth (canvas, natural cotton, or burlap)
- Quality wood glue (optional)

Squaring Jig Plans

On one long side of the 4ft x 1" x 3ft plywood, match the strip of 4ft x 1" x 2" oak board so it overlaps the plywood top by ¼". (You can rout out the oak to snug-fit the plywood for even more support before fastening, but it's not necessary.)

Pre-drill screw holes at intervals to avoid splitting the plywood. Fasten with at least 6 evenly-spaced decking screws to keep the oak secure and square at 90 degrees to the plywood.

The squaring jig will sit on top of any table saw (with its blade retracted) and help as a condensed work surface. If the cuts are square, the boxes will square themselves naturally against the edge during fastening.

To use, stand one long side of a bee box against the fence and build from there.

Master Measurements (using 3/4" Pine)

(Length x Depth x Width)

Hive Box:

2 pieces 13 3/8" x 8 1/4" (long sides)

2 pieces 11 13/16" x 8 1/4" (short ends)

8 pieces per box 12 5/8" x 3/18" x 15/16" (top bars)

Quilt Box:

2 pieces 11 3/16" x 3 15/16" (short sides)

2 pieces 13 3/8" x 3 15/16" (long ends)

1 piece burlap/canvas/cloth 15 3/4" x 15 3/4"

Base Stand/Entrance:

2 pieces 13 3/16" x 6 5/8" (floor)

2 pieces 8 1/4" x 1 3/16" (supports)

1 piece 16 1/8" x 6 5/16" (landing board)

4 pieces 8" x 2.5" x 2.5" (legs)

Roof:

2 pieces 15 3/8" x 8 1/4" (end gables)

2 pieces 13 3/4" x 4 3/4" (sides)

2 pieces 19 11/16" x 8 1/4" (roofing panels)

1 piece 19 11/16" x 2 3/8" (roof ridge)

1 piece 15 3/8" x 13 3/4" exterior plywood (inner cover board)

The Warre Hive Box Cuts

Hive Box:

2 pieces 13 3/8" x 8 ¼" (long sides)

2 pieces 11 13/16" x 8 ¼" (short ends to be routed in next step)

8 pieces per box 12 5/8" x 3/8" x 15/16" (top bars)

Though you will only start with one, it is a good idea to make 3-4 Warre hive boxes to be prepared for the season.

Rout the Warre Hive Box Ends

This is the lip that the top bars rest on within the boxes. All Warre hive boxes have the same measurements for this lip, 3/8" x 3/8".

Using a piece of scrap wood for testing, set your table saw blade and edge guide fence to make a 3/8" deep cut at 3/8" depth into the length-side (11 3/16") of the end board.

Repeat with remaining end boards. Once that first cut is completed, run the end board vertically along the blade to make the second cut. You should now have a 3/8" x 3/8 lip on which the top bars will rest.

Of course, if you have a dado blade, this process becomes simpler.

Build the Warre Boxes

Using the squaring jig, dry fit and clamp the boxes to ensure everything is straight and flush. On the end boards, the routed lip is facing up and inside. The side boards run the length of the outside of each box, sandwiching the end pieces so it fits like this:

When the everything matches up flush and even, fasten the box sides together. Use 3-4 decking screws when joining each edge, making sure to pre-drill to avoid splitting the wood.

Affix handles made from scrap wood to the box sides according to your preference using two (2) #6 1-1/4" steel screws.

To make the top bars themselves, 12 5/8" x 3/8" x 15/16", cut your strips to size, then run a channel in the middle, lengthwise. This can be done a number of ways, but what makes sense for us is to set the table saw blade to 2/16", set the guide fence and use a good push stick to control the cut.

Once the channels are cut, fit an appropriate width strip of wood, equal to your table saw blade width, so that the strip hangs down about ¼ from the surface. Wood glue can be used if required.

This strip does not run end to end, but stops short so that the bars can rest on the Warre box end rabbits, and that there is room for bees to move around both sides.

Build the Warre Quilt Box

Quilt Box:

2 pieces 11 3/16" x 3 15/16" (short sides)

2 pieces 13 3/8" x 3 15/16" (long ends)

1 piece burlap/canvas/cloth 15 ¾" x 15 ¾"

The quilt box in a Warre hive absorbs moisture from the hive through a permeable cloth, like burlap, drawing it into an organic material, like wood shavings. The quilt rests between the top hive box and the roof.

Once you have made the cuts, use the squaring jig and clamps to bring the box together. When it is fitted properly, pre-drill two holes on each joined edge and drive in the decking screws.

Now, drape the 15 ¾" x 15 ¾" cloth tightly over one open end and staple to the top rim. Once tight, staple the overhanging cloth to the sides of the quilt. Trim neatly.

When assembled with a complete hive, the cloth will lie on the bottom above the top hive box, allowing you to fill the quilt loosely with absorbent natural material.

Warre Base Stand/Entrance

Base Stand/Entrance:

2 pieces 13 3/16" x 6 5/8" (floor)

2 pieces 8 ¼" x 1 3/16" (fastened supports)

1 piece 16 1/8" x 6 5/16" (landing board)

4 pieces 8" x 2.5" x 2.5" (legs)

The dimensions of a Warre base stand/entrance are slightly smaller than the hive box outer dimensions.

Warre's design provides greater weight support when compared with a standard Langstroth. With the rotation of the boxes, a screened entrance and mite board is not required for ventilation, allowing for a more solid base. However, if you wish to incorporate these, you can easily adjust the plans to your needs.

To begin, measure out the hive entrance from one of the two pieces of 13 3/16 x 6 5/8" floor boards. Find the center along the long side and measure out a centered rectangle 4" wide x 2" deep. Cut away this waste section for the entrance.

Using the squaring jig, lay the two 13 3/16" x 6 5/8" floor pieces back to back with the entrance facing outward. Attach the two 8 ¼" x 1 3/16" fastened supports inset flush with both sides by about ½", using #6 1-1/4" steel screws

Align the 16 1/8" x 6 5/16" landing board center with the entrance so that it overlaps and is flush with the opposite side of the floor. Fasten the landing board securely with several #6 1-1/4" steel screws, careful not to drill into the empty entrance space.

Attach the 4 8" x 2.5" x 2.5" legs to the underside of the base stand/entrance board using deck screws in pre-drilled holes.

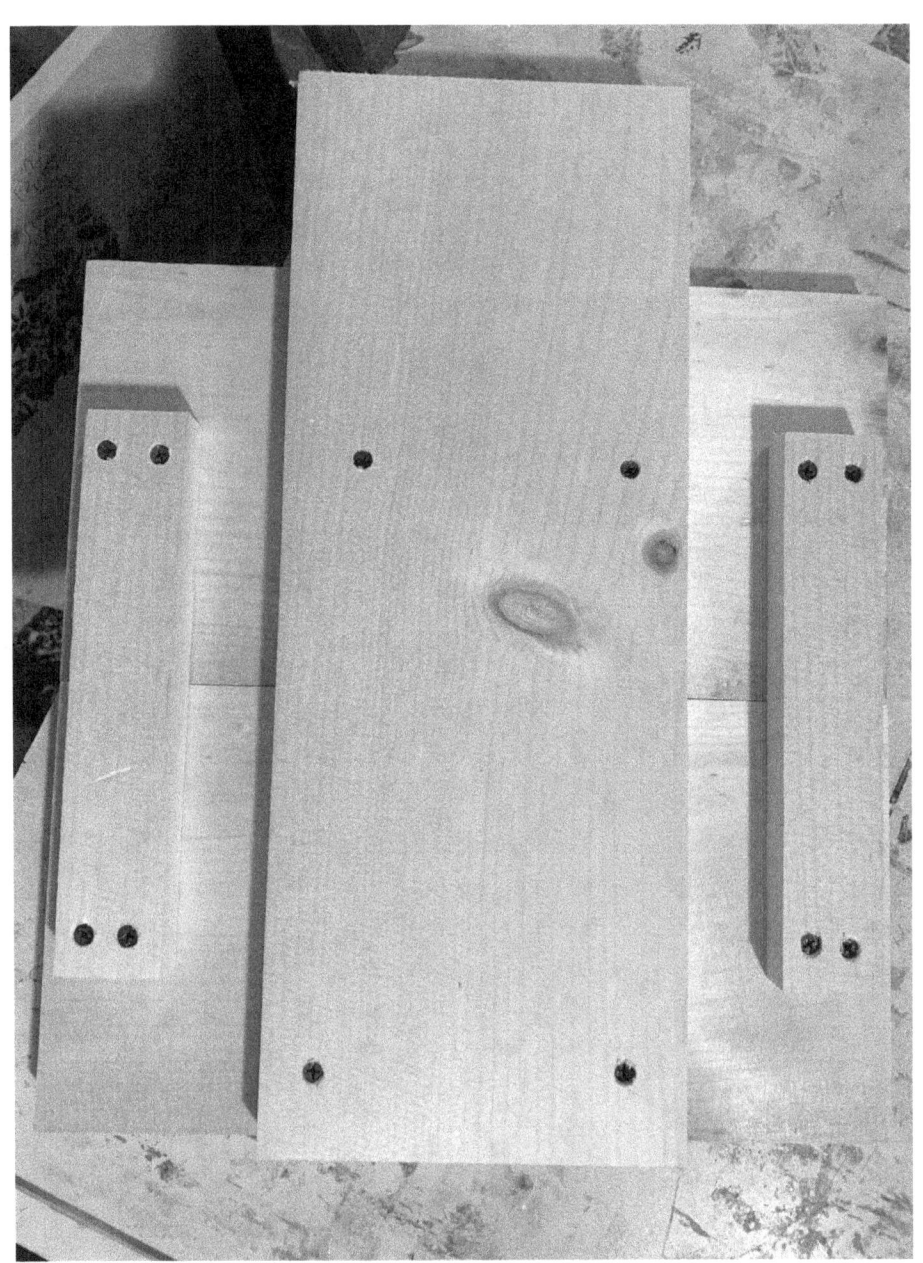

Build the Warre Roof

Roof:

2 pieces 15 3/8" x 8 ¼" (end gables)

2 pieces 13 ¾" x 4 ¾" (sides)

2 pieces 19 11/16" x 8 ¼" (roofing panels)

1 piece 19 11/16" x 2 7/8" (roof ridge)

1 piece 15 3/8" x 13 ¾" exterior plywood (inner cover board)

Start by making the cuts for the end gables. On a flat surface, measure up from each bottom edge 6 5/16" and make a mark. Find the center of the top edge and measure out 13/16" toward each side and make the marks. With a straight edge, connect the top mark to the corresponding side mark.

Cut away the waste to create the roof angle (72 degrees).

Using the squaring jig, matchup the two end gables and the two 13 ¾" x 4 ¾" sides, with the sides resting within the end gables. Clamp, and ensure everything is flush and square. Pre-drill and fasten with decking screws.

Rest the 15 3/8" x 13 ¾" inner cover board flush on top of the side boards, pre-drill and affix with decking screws.

Lay the two 19 11/16" x 8 ¼" roofing panels so that they are square with the gables, overhanging roughly 2 3/16". The top of the roofing panels should align with the flat edge of the gables. Pre-drill and fasten with decking screws.

Lastly, lay the 19 11/16" x 2 7/8" roof ridge piece along the open roof line so that it is square, yet overhangs the sides of the roof panels slightly. Pre-drill and fasten with decking screws.

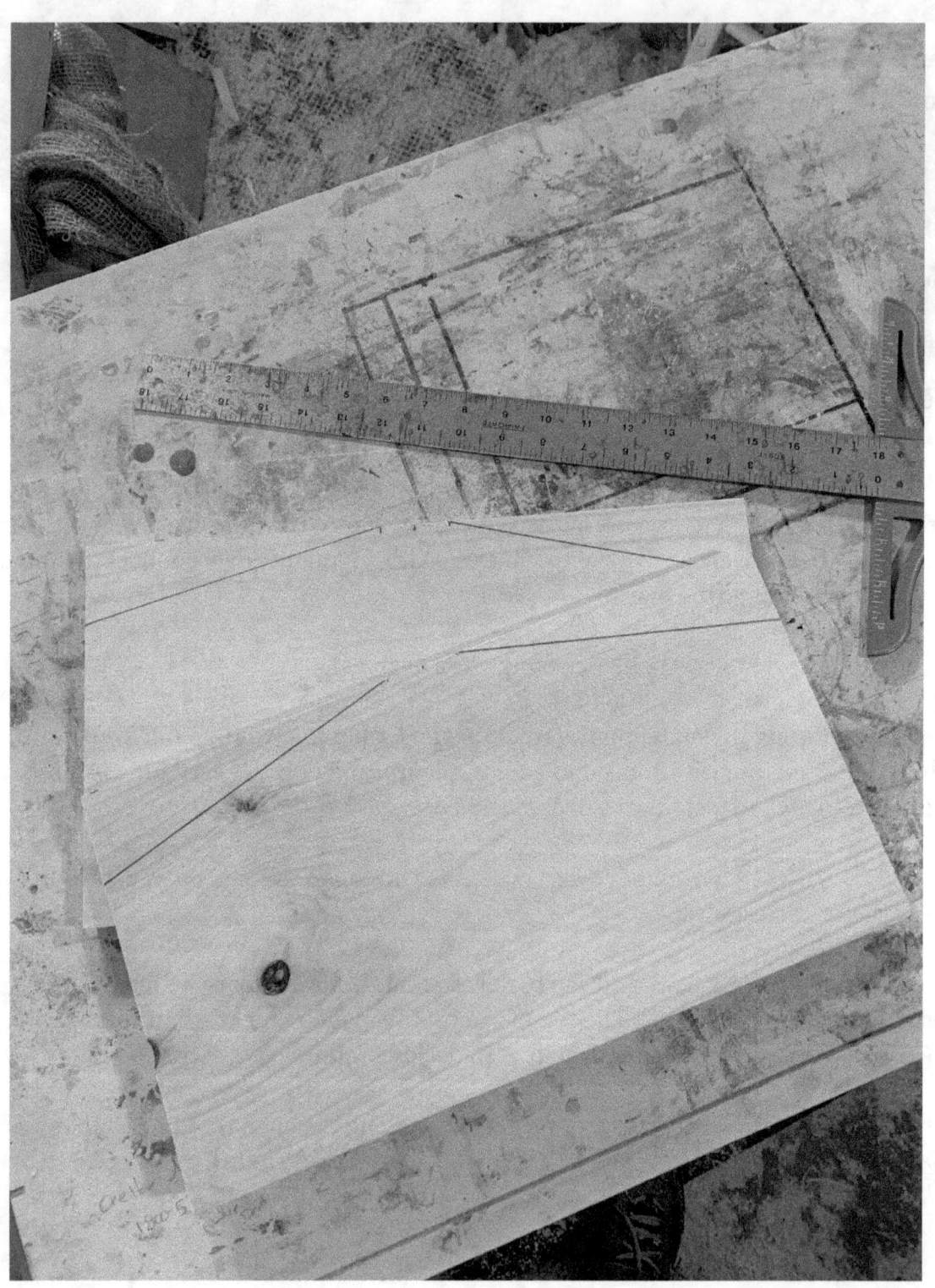

All you need to do now is place the top bars in, protect the exterior with a stain, paint or spar urethane and invite some bees to set up housekeeping.

As you gain confidence with building bee hives, you will discover slight adjustments in measurements, hardware and materials that can make the build more efficient and productive for your particular situation.

Just remember the basics, and have fun!

Best Wishes,

W. Todd Abernathy
New Hampshire WoodWorks

By Abbe Emile Warre

NOTES